Unveiling the Cosmos

Unveiling the Cosmos

Matthew Petchinsky

1

Unveiling the Cosmos: A Guide to Stargazing and Space Exploration

By: Matthew Petchinsky

Introduction: A Comprehensive Guide to Understanding the Night Sky and Space Exploration

For as long as humans have roamed the Earth, we have looked to the night sky in awe and curiosity. From the early stargazers of ancient civilizations to the modern-day astronomers exploring the vast reaches of the universe, the cosmos has been a source of wonder, mystery, and discovery. *Unveiling the Cosmos: A Guide to Stargazing and Space Exploration* is designed to bridge the gap between casual observers and dedicated astronomers, offering a detailed roadmap for understanding the celestial realm and humanity's journey into space.

The Allure of the Night Sky

The night sky is a living, breathing tapestry of celestial objects, each with its own story to tell. The stars twinkle in coded messages of their past, the planets wander through their eternal dance, and cosmic phenomena such as nebulae and galaxies remind us of the universe's vastness. To the untrained eye, these lights may seem like mere dots scattered randomly across a black canvas, but with knowledge and guidance, they become familiar friends—each with a unique history, composition, and role in the cosmos.

Stargazing is more than just an enjoyable pastime; it is an invitation to explore the universe from our own backyards. Whether you're using the naked eye, binoculars, or a telescope, the night sky holds a wealth of secrets waiting to be uncovered. It connects us with the past, reminding us of the navigators who relied on constellations, the astronomers who charted planetary motions, and the scientists who unlocked the physics governing celestial bodies.

This guide will teach you how to decipher the celestial sphere, recognize constellations, track planetary movements, and even observe deep-space objects. It will provide techniques for maximizing your viewing experience, from selecting the best locations and equipment to understanding seasonal changes in the sky.

Humanity's Journey Beyond Earth

While the stars have long fascinated us, humanity's reach has extended beyond mere observation. The last century has been defined by extraordinary advancements in space exploration, from the first artificial satellite, *Sputnik*, to the Apollo moon landings, the Mars rovers, and the ongoing search for exoplanets.

Space exploration is not just the work of astronauts and scientists; it represents the collective curiosity and ambition of humanity. We have sent probes to the outer planets, landed rovers on alien worlds, and developed telescopes that peer into the very birth of the universe. The knowledge gained from these missions reshapes our understanding of physics, biology, and even philosophy, as we ponder our place in the cosmos.

In this book, we will explore the history, achievements, and future of space exploration, covering everything from the first rockets to the possibility of interstellar travel. We will discuss the major players in space exploration—NASA, SpaceX, the European Space Agency (ESA), and others—while also examining the technological breakthroughs that make these missions possible.

What This Book Covers
This comprehensive guide is divided into two key parts:

1. **Stargazing and Observational Astronomy**
    - Understanding the celestial sphere and sky maps
    - The significance of constellations and asterisms
    - The motion of planets and how to track them
    - The phases of the Moon and lunar events
    - Identifying deep-sky objects like galaxies, nebulae, and star clusters
    - The best tools for amateur astronomers, from binoculars to advanced telescopes
    - Light pollution and the importance of dark-sky conservation
2. **Space Exploration and the Future of Astronomy**
    - The history of space travel: From the Space Race to Mars exploration
    - The International Space Station and space habitats
    - The evolution of telescopes: From Galileo to the James Webb Space Telescope
    - The search for extraterrestrial life and habitable planets
    - Space tourism and the commercialization of space travel
    - The role of artificial intelligence and robotics in future missions
    - Theoretical possibilities: Warp drives, interstellar travel, and beyond

Each chapter will provide an accessible yet in-depth exploration of these topics, ensuring that both beginners and seasoned space enthusiasts will find valuable insights.

Why This Book Matters

Space exploration and stargazing are not merely academic pursuits—they represent humanity's deepest yearning to understand our origins and destiny. As we continue to push the boundaries of science and technology, our discoveries will shape the future of our species. The cosmos is not just something to be observed; it is a realm we are destined to explore.

Whether you're an amateur astronomer eager to navigate the night sky or a space enthusiast fascinated by humanity's next frontier, *Unveiling the Cosmos: A Guide to Stargazing and Space Exploration* will equip you with the knowledge and tools to embark on your own cosmic journey.

Look up, wonder, and explore—because the universe is waiting.

Chapter 1: The Basics of Astronomy: Understanding the Universe

1.1 Introduction to Astronomy

Astronomy is one of the oldest sciences, dating back thousands of years to ancient civilizations that gazed up at the night sky in awe and curiosity. Over the centuries, our understanding of the universe has evolved dramatically, moving from mythology and philosophical speculation to a precise, mathematical, and observational science. Today, astronomy is divided into various subfields, each dedicated to studying different aspects of the cosmos, from the motion of planets to the formation of galaxies and the fundamental forces governing the universe.

This chapter lays the foundation for understanding the vast, mysterious, and beautiful cosmos, introducing key concepts and terminology essential for any aspiring astronomer.

1.2 The Structure of the Universe

The universe is a vast, dynamic system composed of countless celestial objects and phenomena. To understand astronomy, one must first grasp the large-scale structure of the cosmos and how its components interact.

1.2.1 The Cosmic Scale

The universe is enormous beyond human comprehension, spanning an estimated 93 billion light-years in observable space. It consists of:

- **Stars** – Massive spheres of plasma that generate energy through nuclear fusion.
- **Planets** – Rocky or gaseous bodies orbiting stars.
- **Moons (Natural Satellites)** – Smaller celestial bodies orbiting planets.
- **Asteroids and Comets** – Remnants from the early solar system, composed of rock, ice, and metals.
- **Galaxies** – Massive systems of stars, gas, dust, and dark matter bound by gravity.
- **Nebulae** – Enormous clouds of gas and dust where stars are born.
- **Black Holes** – Extremely dense points in space with gravitational fields so strong that nothing, not even light, can escape them.

1.2.2 The Hierarchical Structure of the Universe

The cosmos is structured hierarchically:

1. **Planetary Systems** – Like our Solar System, consisting of a central star and its orbiting planets.
2. **Star Clusters** – Groups of stars that share a common origin.
3. **Galaxies** – Collections of billions of stars, such as the Milky Way.
4. **Galaxy Groups and Clusters** – Galaxies bound together by gravity, forming larger cosmic structures.
5. **The Cosmic Web** – The largest known structure, consisting of galaxy superclusters interconnected by filaments of dark matter and gas.

1.3 The Laws of the Universe

The universe operates under a set of fundamental laws that govern the behavior of celestial bodies and cosmic phenomena.

1.3.1 Gravity: The Force That Holds Everything Together

Discovered by Isaac Newton and later refined by Albert Einstein, gravity is the force that governs the motion of planets, stars, and galaxies. It keeps planets in orbit around stars and determines the formation and evolution of the universe.

1.3.2 The Laws of Motion

Newton's three laws of motion explain how celestial objects move:

1. **An object at rest stays at rest, and an object in motion stays in motion unless acted upon by an external force.**
2. **Force equals mass times acceleration (F=ma).**
3. **For every action, there is an equal and opposite reaction.**

These principles are crucial in understanding planetary orbits, rocket launches, and space travel.

1.3.3 The Theory of Relativity

Einstein's Theory of Relativity revolutionized our understanding of space and time. His famous equation, $\mathbf{E=mc^2}$, demonstrates how mass can be converted into energy. General Relativity explains how massive objects warp spacetime, leading to phenomena like black holes and gravitational waves.

1.4 Light and the Electromagnetic Spectrum

Astronomers rely on light to study celestial objects. Since we cannot travel to distant stars and galaxies, we must analyze the light they emit to determine their composition, motion, and distance.

1.4.1 Types of Light in Astronomy

Light exists in a spectrum of different wavelengths:

- **Radio Waves** – Used in radio astronomy to detect pulsars and cosmic background radiation.
- **Microwaves** – Help astronomers study the remnants of the Big Bang.
- **Infrared Light** – Allows us to see through cosmic dust and observe forming stars.
- **Visible Light** – The only part of the spectrum humans can see with the naked eye.
- **Ultraviolet Light** – Used to study hot stars and active galaxies.
- **X-Rays** – Reveal black holes, neutron stars, and supernova remnants.
- **Gamma Rays** – The most energetic waves, used to detect extreme cosmic explosions.

Each type of light provides unique insights into the universe, which is why modern astronomy relies on a variety of telescopes and instruments.

1.4.2 The Doppler Effect and Redshift

As objects in space move, their light shifts in wavelength. When an object moves away from us, its light shifts toward the red end of the spectrum (redshift), and when it moves closer, it shifts toward blue (blueshift). This principle helps astronomers determine the motion of stars and galaxies, providing evidence for the expansion of the universe.

1.5 Time and Space: The Expanding Universe

1.5.1 The Big Bang Theory

The most widely accepted explanation for the origin of the universe is the Big Bang Theory. According to this model, the universe began as an incredibly hot and dense point approximately 13.8 billion years ago and has been expanding ever since.

1.5.2 The Expanding Universe

Observations of distant galaxies show that they are moving away from us, providing strong evidence that space itself is expanding. This discovery, made by Edwin Hubble in the 1920s, led to the formulation of **Hubble's Law**, which states that the farther away a galaxy is, the faster it is receding.

1.5.3 The Ultimate Fate of the Universe

There are three main theories about how the universe will end:

1. **The Big Freeze** – The universe continues expanding until all energy is depleted, leading to a cold, lifeless cosmos.
2. **The Big Crunch** – The universe's expansion reverses, causing everything to collapse into a singularity.
3. **The Big Rip** – Dark energy accelerates expansion until galaxies, stars, and even atoms are torn apart.

Current evidence suggests that the universe will continue expanding indefinitely, leading to a slow cooling over trillions of years.

1.6 Conclusion: Why Astronomy Matters

Understanding the universe is not just about scientific knowledge—it is about our place in the cosmos. Astronomy helps us answer fundamental questions: Where did we come from? Are we alone? What is the ultimate fate of the universe?

Through telescopes, satellites, and spacecraft, we continue to push the boundaries of our knowledge, unlocking the secrets of the stars and beyond. Whether you are an amateur stargazer or a professional scientist, the night sky offers an infinite realm of exploration and discovery.

This chapter has provided a foundational understanding of astronomy, preparing you for deeper explorations into the wonders of the universe. As we move forward in this book, we will examine the tools of stargazing, the history of space exploration, and the groundbreaking discoveries that have shaped our understanding of the cosmos.

Look up, wonder, and begin your journey through the universe.

Chapter 2: The Constellations: Mythology and Science

2.1 Introduction to Constellations

For millennia, humans have looked to the night sky and identified patterns among the stars. These groupings, known as **constellations**, have played a crucial role in mythology, navigation, and scientific study. While early civilizations associated them with gods, heroes, and creatures of legend, modern astronomy uses them as a celestial coordinate system to locate stars, planets, and other cosmic phenomena.

This chapter explores the dual nature of constellations—their mythological roots and their scientific significance—offering insights into how they have shaped human understanding of the cosmos.

2.2 The Origins of Constellations: Myths and Legends

The concept of constellations dates back thousands of years, with early civilizations using them to tell stories, mark the changing seasons, and guide travelers. Many cultures independently developed their own celestial narratives, but some constellations have remarkably similar interpretations across different societies.

2.2.1 Constellations in Ancient Civilizations

Babylonians and Sumerians (3000 BCE - 500 BCE)

The Babylonians were among the first recorded cultures to document star groupings, laying the foundation for the Western zodiac. They divided the sky into 12 sections, each associated with a particular constellation along the **ecliptic** (the Sun's apparent path across the sky). These became the zodiac signs we recognize today.

Egyptians (2500 BCE - 500 BCE)

The ancient Egyptians linked their constellations to their gods, such as Osiris (associated with **Orion**) and Horus (linked to **the Sun and Sirius**). They used stars like **Sirius** to predict the annual flooding of the Nile, which was crucial for agriculture.

Greeks (800 BCE - 300 CE)

Greek mythology gave us many of the constellations still in use today. The **42 Greek constellations** documented by **Claudius Ptolemy** in his *Almagest* (2nd century CE) formed the foundation of modern Western star charts. These include famous stories such as **Orion the Hunter, Pegasus the Winged Horse, and Andromeda the Chained Princess**.

Chinese Astronomy (2000 BCE - Present)

Unlike the Western system, ancient Chinese astronomy divided the sky into **28 lunar mansions** rather than constellations, with each segment representing different mythological creatures. The **Four Celestial Beasts**—Azure Dragon (spring), White Tiger (autumn), Vermilion Bird (summer), and Black Tortoise (winter)—were key figures in Chinese star lore.

Indigenous and Other Cultural Interpretations

- **The Australian Aboriginals** saw constellations as animals and ancestral spirits. The "**Emu in the Sky**" (which is not made of stars but dark nebulae) is a prime example.
- **The Navajo** recognized star formations as part of their spiritual landscape, with constellations like **First Man and First Woman** shaping creation myths.
- **Hindu Astronomy** developed **Nakshatras**, 27 lunar mansions that were vital for astrology and timekeeping.

These myths were not just stories—they were practical tools for tracking time, seasons, and navigation.

2.3 The Modern Constellations

Today, the International Astronomical Union (IAU) officially recognizes **88 constellations**, covering both the Northern and Southern Hemispheres. These serve as a celestial map, helping astronomers locate celestial objects more easily.

2.3.1 How Constellations Are Defined

Unlike ancient times, when constellations were loosely defined based on visible star patterns, modern astronomy assigns each constellation a specific region of the sky. This means even if a star doesn't fit within an obvious pattern, it still belongs to a constellation based on celestial coordinates.

The **celestial sphere** is an imaginary sphere surrounding Earth, onto which all stars and celestial bodies appear to be projected. To navigate this sphere, astronomers use a system similar to latitude and longitude, with **right ascension (RA)** and **declination (Dec)** to pinpoint locations.

2.4 Notable Constellations and Their Stars

Certain constellations are more prominent due to their bright stars, unique patterns, or significance in both mythology and science. Below are some of the most well-known constellations.

2.4.1 The Zodiac Constellations

The **12 zodiac constellations** are those along the ecliptic, the path the Sun appears to travel throughout the year:

1. **Aries (The Ram)** – Associated with the golden ram of Greek mythology.
2. **Taurus (The Bull)** – Contains **Aldebaran** and the famous **Pleiades cluster**.
3. **Gemini (The Twins)** – Home to **Castor and Pollux**, two bright stars.
4. **Cancer (The Crab)** – Hosts the **Beehive Cluster**.
5. **Leo (The Lion)** – Dominated by **Regulus**, one of the brightest stars in the sky.
6. **Virgo (The Maiden)** – The location of **Spica** and part of the massive Virgo Galaxy Cluster.
7. **Libra (The Scales)** – One of the few constellations representing an inanimate object.
8. **Scorpius (The Scorpion)** – Features **Antares**, a bright red supergiant star.
9. **Sagittarius (The Archer)** – Home to the center of the Milky Way and numerous deep-sky objects.
10. **Capricornus (The Sea Goat)** – A faint constellation tied to ancient mythology.
11. **Aquarius (The Water Bearer)** – Linked to the legend of Ganymede, Zeus's cupbearer.
12. **Pisces (The Fish)** – Represents two fish tied together, a story from Greek mythology.

2.4.2 The Most Recognizable Constellations

- **Orion (The Hunter)** – One of the most famous constellations, featuring **Betelgeuse (a red supergiant) and Rigel (a blue supergiant)**.
- **Ursa Major (The Great Bear)** – Home to the **Big Dipper**, a famous asterism used for navigation.
- **Ursa Minor (The Little Bear)** – Contains **Polaris, the North Star**, an essential guide for navigation.
- **Cassiopeia (The Queen)** – Recognizable by its distinctive **W** shape.
- **Cygnus (The Swan)** – Contains **Deneb**, one of the brightest stars in the sky.
- **Canis Major (The Great Dog)** – Home to **Sirius, the brightest star in the night sky**.

2.5 How Constellations Aid Astronomy

2.5.1 Constellations as Astronomical Landmarks

Astronomers use constellations to locate objects such as nebulae, star clusters, and galaxies. Since deep-sky objects are often faint, identifying them by their neighboring constellations makes tracking them easier.

2.5.2 Seasonal Visibility of Constellations

Different constellations are visible at different times of the year due to Earth's orbit around the Sun:

- **Winter Constellations** – Orion, Taurus, Gemini
- **Spring Constellations** – Leo, Virgo, Bootes
- **Summer Constellations** – Cygnus, Scorpius, Sagittarius
- **Autumn Constellations** – Andromeda, Pegasus, Aquarius

2.5.3 Constellations and Deep-Sky Objects

Some constellations contain famous deep-sky objects, including:

- **Orion Nebula (M42) in Orion** – A stellar nursery.
- **Andromeda Galaxy (M31) in Andromeda** – The closest spiral galaxy to our own.
- **The Pleiades in Taurus** – One of the nearest and brightest open star clusters.

2.6 Conclusion

Constellations serve as both scientific tools and cultural relics, connecting us to the knowledge and stories of our ancestors while guiding modern astronomy. Whether you're gazing up at Orion's Belt, tracing the Big Dipper, or locating the Andromeda Galaxy, understanding constellations enriches the experience of stargazing and deepens our appreciation of the universe.

Look up, explore, and continue the journey through the stars.

Chapter 3: Observing Planets and Moons

3.1 Introduction to Planetary Observation

Since the dawn of astronomy, humans have been captivated by the wandering lights in the night sky—celestial objects that moved against the fixed backdrop of stars. These "wanderers" were later identified as planets, members of our Solar System that orbit the Sun just as Earth does. Observing planets and their moons has not only provided a glimpse into the workings of our own cosmic neighborhood but has also revealed stunning geological activity, atmospheric wonders, and even the potential for extraterrestrial life.

This chapter is dedicated to guiding you through the methods, tools, and techniques required to observe planets and moons, detailing what to expect when viewing each one and offering insights into the unique characteristics of these fascinating celestial bodies.

3.2 The Best Times and Conditions for Observing Planets

Unlike stars, which shine due to nuclear fusion, planets shine by reflecting sunlight. Because of their proximity to Earth, they follow predictable paths, making it easier to determine when they are best observed.

3.2.1 Key Factors for Planetary Observation

- **Opposition** – When a planet is opposite the Sun in the sky, it is closest to Earth and fully illuminated. This is the best time to observe outer planets like Mars, Jupiter, and Saturn.
- **Conjunction** – When a planet is on the far side of the Sun, making it difficult to observe.
- **Elongation** – The angle between the Sun and an inner planet (Mercury or Venus). Greatest elongation provides the best viewing conditions for these planets.
- **Seeing Conditions** – The steadiness of Earth's atmosphere affects the clarity of planetary details. Viewing is best when the air is calm and stable.
- **Light Pollution** – Observing from a dark-sky location enhances visibility, though planets are often bright enough to be seen even in urban areas.

3.3 Observing the Planets of Our Solar System

Each planet has its own distinct characteristics that make observing it a unique experience.

3.3.1 Mercury: The Elusive Inner World

- **Best Time to Observe**: At greatest elongation, shortly after sunset or before sunrise.
- **What to Expect**: A tiny, bright disk that undergoes phases like the Moon.
- **Observation Tips**: Due to its proximity to the Sun, Mercury is always seen near the horizon, requiring a clear, unobstructed view.

3.3.2 Venus: Earth's Twin in a Fiery Veil

- **Best Time to Observe**: At greatest elongation, either in the morning or evening.
- **What to Expect**: A brilliant white disk, sometimes crescent-shaped due to its phases.
- **Observation Tips**: Venus is one of the brightest objects in the sky, often mistaken for a UFO. Through a telescope, you can see its phases, but thick clouds obscure its surface.

3.3.3 Mars: The Red Planet

- **Best Time to Observe**: At opposition, every 26 months.
- **What to Expect**: A small reddish-orange disk, with surface features visible during ideal conditions.
- **Observation Tips**: With a medium to large telescope, you can observe **polar ice caps**, dark surface markings, and even dust storms.

3.3.4 Jupiter: The Giant of the Solar System

- **Best Time to Observe**: At opposition, visible for much of the year.
- **What to Expect**: A large, striped disk with cloud bands and the Great Red Spot.
- **Observation Tips**: Even with binoculars, you can see **Jupiter's four largest moons (Io, Europa, Ganymede, Callisto)** as tiny points of light moving around the planet. A telescope reveals the planet's famous **cloud bands** and possibly the Great Red Spot.

3.3.5 Saturn: The Lord of the Rings

- **Best Time to Observe**: At opposition, visible for much of the year.
- **What to Expect**: A stunning golden-yellow disk with its famous ring system.
- **Observation Tips**: Even a small telescope will reveal **Saturn's rings**. With a larger telescope, you can spot the **Cassini Division** (a dark gap in the rings) and some of its **many moons**, including **Titan**.

3.3.6 Uranus: The Ice Giant

- **Best Time to Observe**: At opposition.
- **What to Expect**: A small, faint, bluish-green disk.
- **Observation Tips**: Uranus is just visible to the naked eye under dark skies. A telescope reveals its smooth disk, but surface details are difficult to see.

3.3.7 Neptune: The Distant Blue World

- **Best Time to Observe**: At opposition.
- **What to Expect**: A tiny, deep-blue disk.
- **Observation Tips**: Neptune is only visible through binoculars or a telescope. A large telescope may reveal its **moon Triton**, which has an atmosphere and active geysers.

3.4 Observing Moons in the Solar System

While planets are exciting, their moons are equally fascinating. Some are more active and geologically diverse than any planet.

3.4.1 Observing Earth's Moon

- **Phases**: The Moon goes through predictable **phases** (New, Crescent, First Quarter, Gibbous, Full, Last Quarter).
- **Best Time to Observe**: The first quarter and waxing gibbous phases provide the best shadows for detail.
- **Lunar Features**:
    - **Craters**: Impact sites such as **Tycho and Copernicus**.
    - **Maria**: Dark, flat plains (ancient lava flows).
    - **Mountain Ranges**: The **Apennine Mountains** are a prominent lunar feature.

3.4.2 The Galilean Moons of Jupiter

- **Io**: The most volcanically active body in the Solar System.
- **Europa**: A smooth, icy world with a subsurface ocean.
- **Ganymede**: The largest moon in the Solar System, larger than Mercury.
- **Callisto**: A heavily cratered moon with an ancient surface.

3.4.3 Saturn's Moons

- **Titan**: The second-largest moon in the Solar System, with a thick atmosphere and methane lakes.
- **Enceladus**: A small moon with geysers spraying water ice into space.

3.4.4 Notable Moons of Other Planets

- **Triton (Neptune)**: Has retrograde motion and active geysers.
- **Phobos & Deimos (Mars)**: Small, irregularly shaped moons likely captured asteroids.

3.5 Tools and Techniques for Observing Planets and Moons

- **Binoculars**: A good starting point for observing planets, the Moon, and Jupiter's moons.
- **Telescopes**:
    - **Small telescopes (60-100mm aperture)**: Good for observing the Moon, Venus, Mars, Jupiter, and Saturn.
    - **Medium telescopes (100-200mm aperture)**: Reveal cloud bands, planetary moons, and lunar details.
    - **Large telescopes (200mm+ aperture)**: Needed for Neptune, Uranus, and faint details on other planets.
- **Filters**:
    - **Neutral-density filters** help reduce brightness for lunar observation.
    - **Colored filters** can enhance planetary details (e.g., red for Mars, blue for Jupiter).

3.6 Conclusion

Observing planets and their moons is one of the most rewarding aspects of astronomy. Whether you're tracking Jupiter's moons, marveling at Saturn's rings, or watching the phases of Venus, each celestial body tells a story of cosmic history and evolution. With patience, the right tools, and an understanding of when and how to observe, you can embark on a journey to explore our Solar System from your own backyard.

The universe is at your fingertips—just look up.

Chapter 4: Meteor Showers and Celestial Events

4.1 Introduction to Meteor Showers and Celestial Events

The night sky is not only home to distant stars, planets, and galaxies—it is also a dynamic theater of motion and change. Among the most spectacular events visible to the naked eye are **meteor showers**, which occur when Earth passes through the debris left behind by comets and asteroids. Other celestial phenomena, such as eclipses, planetary conjunctions, and auroras, add to the wonder of the cosmos, giving stargazers a reason to look up throughout the year.

This chapter explores the science behind meteor showers, their connection to comets and asteroids, and a range of other celestial events that light up the sky.

4.2 Understanding Meteor Showers

4.2.1 What is a Meteor Shower?

Meteor showers occur when **Earth passes through the debris field of a comet or asteroid**. The tiny fragments, often no larger than a grain of sand, burn up upon entering Earth's atmosphere, creating bright streaks of light known as **meteors**. If a meteor survives the atmospheric entry and reaches the ground, it is called a **meteorite**.

The **point in the sky from which meteors appear to radiate** is called the **radiant**, and it is typically named after the constellation where it is located.

4.2.2 The Life Cycle of a Meteor Shower

1. **Cometary Origins** – When a comet orbits the Sun, its ice and dust begin to vaporize, leaving behind a trail of debris.
2. **Earth's Intersection** – As Earth moves along its orbit, it occasionally crosses the debris path.
3. **Atmospheric Entry** – The particles collide with Earth's atmosphere at high speeds (up to **160,000 mph**), causing frictional heating that produces the bright streaks of light we see.

4.2.3 The Difference Between Meteors, Meteorites, and Fireballs

- **Meteors** – Small space debris burning up in Earth's atmosphere.
- **Meteorites** – Large fragments that survive the fall and reach Earth's surface.
- **Fireballs** – Exceptionally bright meteors that can last several seconds and may explode in the sky (bolides).

4.3 Major Meteor Showers Throughout the Year

Certain meteor showers occur annually when Earth revisits specific debris trails. Below are the most famous and active meteor showers.

4.3.1 The Quadrantids (January 3-4)

- **Parent Body**: Likely from asteroid 2003 EH1.
- **Peak Rate**: Up to **120 meteors per hour** under ideal conditions.
- **Best Viewing**: Northern Hemisphere, just before dawn.

4.3.2 The Lyrids (April 21-22)

- **Parent Body**: Comet C/1861 G1 (Thatcher).
- **Peak Rate**: Around **20 meteors per hour**.
- **Best Viewing**: Late night into early morning.

4.3.3 The Eta Aquariids (May 5-6)

- **Parent Body**: **Halley's Comet**.
- **Peak Rate**: 50+ meteors per hour in the Southern Hemisphere.
- **Best Viewing**: Early morning, best seen in the tropics and Southern Hemisphere.

4.3.4 The Delta Aquariids (July 28-29)

- **Parent Body**: Uncertain (possibly Comet 96P/Machholz).
- **Peak Rate**: 20 meteors per hour.
- **Best Viewing**: Midnight to dawn.

4.3.5 The Perseids (August 11-13)

- **Parent Body**: Comet 109P/Swift-Tuttle.
- **Peak Rate**: 60-100 meteors per hour.
- **Best Viewing**: One of the most reliable and brightest showers, best seen after midnight.

4.3.6 The Orionids (October 21-22)

- **Parent Body**: **Halley's Comet**.
- **Peak Rate**: Around 20 meteors per hour.
- **Best Viewing**: Best seen in the pre-dawn hours.

4.3.7 The Leonids (November 17-18)

- **Parent Body**: Comet 55P/Tempel-Tuttle.
- **Peak Rate**: Normally 15 meteors per hour, but can produce **meteor storms** (up to 1000 meteors per hour) every 33 years.
- **Best Viewing**: Late night to early morning.

4.3.8 The Geminids (December 13-14)

- **Parent Body**: **3200 Phaethon** (an asteroid rather than a comet).
- **Peak Rate**: 100-120 meteors per hour.
- **Best Viewing**: All night long; produces some of the brightest meteors.

4.4 Other Celestial Events

Meteor showers are just one type of sky event. Many other spectacular occurrences can be seen throughout the year.

4.4.1 Solar and Lunar Eclipses

Eclipses occur when celestial bodies align perfectly, causing one to cast a shadow on another.

- **Solar Eclipse** – Occurs when the **Moon passes in front of the Sun**.
- **Lunar Eclipse** – Occurs when the **Earth's shadow falls on the Moon**.

4.4.2 Planetary Conjunctions

A **planetary conjunction** happens when two or more planets appear very close together in the sky. Some notable conjunctions include:

- **The Great Conjunction of Jupiter and Saturn (2020)** – The closest since 1623, appearing as a "Christmas Star."
- **Venus and Jupiter Conjunctions** – Occur frequently and are easily visible.

4.4.3 Aurora Borealis and Aurora Australis

Auroras, also known as the **Northern and Southern Lights**, occur when **charged particles from the Sun interact with Earth's magnetic field**, causing colorful displays in polar regions.

- **Best Locations**: Canada, Alaska, Scandinavia, and Antarctica.
- **Colors**: Green (oxygen), Red (high-altitude oxygen), Blue/Purple (nitrogen).

4.4.4 Supermoons

A **Supermoon** occurs when the **Moon is at its closest point to Earth (perigee)** while full, making it appear larger and brighter.

4.4.5 Blue Moons

A **Blue Moon** is the second full moon within a single month—a rare event that happens about every **2.5 years**.

4.4.6 Occultations

An **occultation** occurs when one celestial body passes directly in front of another, temporarily obscuring it. Examples include:

- The **Moon passing in front of Mars or Jupiter**.
- An **asteroid blocking a distant star**.

4.5 How to Observe Celestial Events

4.5.1 Essential Equipment for Observation

- **Naked Eye**: Many meteor showers, eclipses, and conjunctions can be observed without any equipment.
- **Binoculars**: Useful for lunar eclipses and auroras.
- **Telescopes**: Helpful for detailed planetary conjunctions and solar observations (with filters).

4.5.2 Ideal Viewing Conditions

- **Dark-Sky Locations**: The farther from city lights, the better.
- **Clear Skies**: Check weather forecasts for clear nights.
- **Moon Phases**: A bright Moon can wash out meteor showers—try observing during a new Moon.

4.6 Conclusion

Meteor showers and celestial events are among the most awe-inspiring phenomena in astronomy. Whether watching a blazing meteor streak across the sky, witnessing an eclipse, or seeing planets align, these events connect us to the greater workings of the cosmos. With the right knowledge, patience, and preparation, you can experience some of the universe's most breathtaking spectacles right from your own backyard.

Look up, wonder, and let the sky guide you.

Chapter 5: The Future of Space Travel and Colonization

5.1 Introduction: Humanity's Next Frontier

For centuries, humans have gazed at the stars and wondered what lies beyond. The dream of venturing beyond Earth, once confined to mythology and science fiction, is now an active pursuit by scientists, engineers, and visionaries. The future of space travel holds the promise of **colonizing other planets, establishing permanent settlements on the Moon and Mars, and exploring the far reaches of the universe**.

With advances in **rocket technology, artificial intelligence, robotics, and sustainable life support systems**, humanity stands at the threshold of an interplanetary future. This chapter delves into the **emerging technologies, challenges, and prospects** of space travel and colonization, offering a glimpse into what the next century may hold.

5.2 The Evolution of Space Travel

5.2.1 Early Spaceflight and the Space Race

Space travel began in the mid-20th century with the **Cold War-driven Space Race** between the United States and the Soviet Union. Key milestones include:

- **Sputnik 1 (1957)** – The first artificial satellite, launched by the Soviet Union.
- **Vostok 1 (1961)** – Yuri Gagarin becomes the first human in space.
- **Apollo 11 (1969)** – The first Moon landing by Neil Armstrong and Buzz Aldrin.
- **Skylab and Mir (1970s-1990s)** – The first space stations.
- **The Space Shuttle Era (1981-2011)** – Reusable spacecraft made space travel more routine.

5.2.2 The Rise of Private Spaceflight

Since the early 2000s, private companies have transformed the space industry. Companies such as **SpaceX, Blue Origin, and Rocket Lab** have reduced launch costs and pioneered new technology, making space more accessible.

Notable achievements:

- **SpaceX's Falcon 9 (2015-present)** – First reusable rocket.
- **SpaceX's Crew Dragon (2020)** – First private company to send astronauts to the ISS.
- **Blue Origin's New Shepard (2021)** – First commercial suborbital space tourism flights.
- **NASA's Artemis Program (2022-Present)** – Aims to return humans to the Moon and establish a lunar base.

5.3 The Next Generation of Space Travel Technologies

Future space exploration depends on breakthroughs in **propulsion, life support, and sustainable energy**.

5.3.1 Advanced Propulsion Systems

Traditional chemical rockets are inefficient for interplanetary travel, prompting research into **faster and more sustainable propulsion methods**:

- **Nuclear Thermal Propulsion (NTP)** – Uses nuclear fission to heat propellant, cutting travel time to Mars in half.
- **Ion Propulsion** – Uses charged particles for sustained, efficient thrust, already used on deep-space probes.
- **Plasma Rockets (VASIMR)** – A high-powered electric propulsion system under development.
- **Antimatter Propulsion** – Theoretical technology that could enable near-light-speed travel.

5.3.2 Space Elevators and Launch Systems

- **Space Elevators** – A cable extending from Earth to orbit, allowing cargo transport without rockets.
- **Starship (SpaceX)** – A fully reusable spacecraft designed for Moon and Mars missions.
- **Railgun and Mass Drivers** – Theoretical systems to launch payloads into space without traditional rockets.

5.4 The Moon: The First Step Toward Colonization

5.4.1 Why the Moon?

The Moon is **our closest celestial body (384,400 km away)** and offers several advantages for space colonization:

- **Lower Gravity (1/6 of Earth's)** – Easier for launching deep-space missions.
- **Water Ice in Polar Craters** – Can be used for drinking water and fuel production.
- **Raw Materials** – Contains helium-3, a potential energy source for nuclear fusion.

5.4.2 Lunar Colonization Plans

1. **NASA Artemis Base Camp (2020s-2030s)** – Permanent lunar habitat at the Moon's south pole.
2. **China's Lunar Research Station (2030s)** – Planned joint project with Russia.
3. **ESA's Moon Village (2040s)** – A proposed international base using **3D printing technology** to build habitats from lunar soil.

Challenges:

- **Radiation exposure** – No atmosphere to block cosmic rays.
- **Extreme temperatures** – Ranges from -250°F to 250°F (-157°C to 121°C).
- **Lunar dust issues** – Abrasive and hazardous to equipment.

5.5 Mars: The Next Frontier

5.5.1 Why Colonize Mars?

- **Has an atmosphere (thin, mostly CO_2).**
- **Possesses frozen water at the poles.**
- **A day on Mars (24.6 hours) is similar to Earth's.**
- **Potential for terraforming in the distant future.**

5.5.2 Mars Colonization Plans

1. **SpaceX's Mars Mission (Starship Program)** – Aims to send humans to Mars by the 2030s.
2. **NASA's Mars Plan** – Focuses on robotic missions before a human presence.
3. **Mars Society's Mars Direct Plan** – A simplified, cost-effective method for colonization.

Challenges:

- **Long travel time (6-9 months one way).**
- **Harsh environment (dust storms, radiation).**
- **Lack of oxygen (requires life-support systems).**

Solutions:

- **Underground habitats** – Protection from radiation.
- **Hydroponic and aeroponic farming** – Growing food without soil.
- **Nuclear reactors** – Reliable energy source.

5.6 The Future of Interstellar Travel

Beyond the Moon and Mars, humanity dreams of traveling to **other star systems**. Current ideas include:

5.6.1 The Breakthrough Starshot Initiative

- Uses **light-powered nanocrafts** to reach **Alpha Centauri** within a few decades.

5.6.2 Theoretical Faster-than-Light Travel

- **Alcubierre Warp Drive** – A concept that could theoretically bend space-time for faster-than-light travel.
- **Wormholes** – Hypothetical shortcuts through space-time.

5.6.3 Cryogenic and Generation Ships

- **Cryogenic sleep** – Putting astronauts in suspended animation for long journeys.
- **Generation ships** – Spacecraft where multiple generations of humans live and die while traveling to distant worlds.

5.7 Ethical and Social Considerations of Space Colonization

5.7.1 Who Owns Space?

- The **Outer Space Treaty (1967)** states that no country can claim celestial bodies. However, private companies are now challenging this idea.

5.7.2 The Impact on Human Evolution

- Long-term space colonization could lead to **genetic changes** due to different gravity and radiation exposure.

5.7.3 Environmental Responsibility

- Space colonization must consider **preserving alien ecosystems** and avoiding contamination.

5.8 Conclusion: The Dawn of a New Space Age

The future of space travel and colonization is no longer confined to science fiction—it is becoming an **inevitable part of human evolution**. The coming decades will see **permanent lunar bases, the first humans on Mars, and the early steps toward interstellar exploration**.

While challenges remain, the potential rewards—**new resources, scientific breakthroughs, and ensuring the survival of humanity**—are worth the effort. Our destiny lies among the stars, and the journey has already begun.

Look up—our future is in the cosmos.

Chapter 6: Black Holes, Wormholes, and the Unknown

6.1 Introduction: The Mysteries of the Cosmos

The universe is a vast, enigmatic expanse filled with celestial wonders that challenge our understanding of physics and reality itself. Among these, **black holes, wormholes, and other unknown cosmic phenomena** stand as some of the most fascinating and perplexing concepts in modern astrophysics.

Black holes—regions of space where gravity is so strong that nothing, not even light, can escape—warp the very fabric of spacetime. Wormholes, theoretical tunnels through spacetime, hint at the possibility of shortcuts across the cosmos. Meanwhile, countless unknown cosmic phenomena continue to puzzle scientists, offering glimpses into **hidden forces and dimensions** beyond our current understanding.

In this chapter, we will explore the science, theories, and ongoing mysteries surrounding black holes, wormholes, and the vast unknown of the universe.

6.2 The Science of Black Holes

6.2.1 What is a Black Hole?

A **black hole** is a region of space where the gravitational pull is so intense that nothing, not even light, can escape. It is formed when a massive star collapses under its own gravity at the end of its life cycle.

The defining boundary of a black hole is called the **event horizon**—the point of no return. Anything crossing this boundary is irreversibly pulled toward the singularity at the center.

6.2.2 Types of Black Holes

There are three main types of black holes:

1. **Stellar-Mass Black Holes**
    - Formed from the collapse of massive stars.
    - Typically range from **3 to 100 times the mass of the Sun**.
2. **Supermassive Black Holes**
    - Found at the centers of galaxies, including our Milky Way.
    - Can be **millions to billions of times the Sun's mass**.
    - Example: **Sagittarius A*** at the heart of our galaxy.
3. **Intermediate-Mass Black Holes**
    - A theoretical class of black holes between stellar-mass and supermassive black holes.
    - Their existence is still debated, but observations suggest they may exist in dense star clusters.
4. **Primordial Black Holes**
    - Hypothetical black holes formed shortly after the Big Bang.
    - Could be tiny, possibly smaller than an atom, but with extreme densities.

6.2.3 How Black Holes Form

Black holes are the end result of massive stars that have exhausted their nuclear fuel. Their formation follows these steps:

1. **Nuclear Fusion Ends** – A star's energy production stops, causing gravitational collapse.
2. **Supernova Explosion** – If the star is large enough, it undergoes a violent explosion.
3. **Collapse into a Black Hole** – If the remaining core is more than about **three solar masses**, gravity crushes it into a singularity.

6.3 The Strange Physics of Black Holes

6.3.1 The Event Horizon and the Singularity

- The **event horizon** is the boundary beyond which escape is impossible.
- The **singularity** is the theoretical point at the black hole's center where density becomes **infinite** and the laws of physics break down.

6.3.2 Time Dilation and Spaghettification

Einstein's **General Theory of Relativity** predicts strange effects near black holes:

- **Time Dilation** – Time slows down near a black hole due to extreme gravity. If you watched a spaceship fall into a black hole, it would appear to **slow down and freeze** at the event horizon.
- **Spaghettification** – The gravitational pull near the event horizon is so strong that it **stretches objects into long, thin shapes** before crushing them.

6.3.3 Hawking Radiation: Can Black Holes Evaporate?

Physicist **Stephen Hawking** proposed that black holes are not entirely black. Due to quantum effects near the event horizon, black holes slowly emit radiation, called **Hawking radiation**, leading to their gradual **evaporation over trillions of years**.

6.4 Wormholes: Cosmic Shortcuts or Science Fiction?

6.4.1 What is a Wormhole?

A **wormhole** is a theoretical tunnel through spacetime, possibly connecting distant points in the universe or even different universes.

- They are based on Einstein's equations and are also called **Einstein-Rosen Bridges**.
- Unlike black holes, which trap everything, wormholes **could allow for faster-than-light travel**.

6.4.2 The Structure of a Wormhole

A traversable wormhole would have:

- **Two mouths (entry and exit points)**.
- **A tunnel-like throat connecting them**.
- **Exotic matter** (with negative energy) to keep it stable and prevent collapse.

6.4.3 Are Wormholes Real?

While wormholes exist in **theory**, there is **no direct observational evidence** yet. However, scientists have proposed ways to detect them, such as looking for gravitational lensing effects that differ from those caused by black holes.

6.4.4 Could We Travel Through a Wormhole?

If wormholes exist, they could allow:

- **Intergalactic Travel** – Jumping across vast distances instantly.
- **Time Travel** – If a wormhole connects different points in time, it could be used as a time machine.
- **Parallel Universes** – Some theories suggest wormholes may link to other dimensions or realities.

6.5 The Unknown: Cosmic Mysteries Yet to Be Solved

6.5.1 Dark Matter and Dark Energy

- **Dark Matter** – Makes up **85% of the universe's matter**, yet it does not emit or absorb light. Its presence is inferred from **gravitational effects** on galaxies.
- **Dark Energy** – The force causing the universe's expansion to accelerate. Scientists estimate that **68% of the universe** is dark energy, yet its true nature remains a mystery.

6.5.2 The Multiverse Theory

Some physicists propose that our universe is just **one of many universes** in a vast **multiverse.** If true, it raises mind-bending questions:

- Are there alternate versions of ourselves?
- Could black holes be gateways to other universes?

6.5.3 The Information Paradox

According to quantum mechanics, **information cannot be lost.** However, black holes seem to destroy anything that falls in, creating a paradox. Recent studies suggest that **information may be encoded on the event horizon** in a way similar to a hologram.

6.6 The Future of Black Hole and Wormhole Research

6.6.1 Imaging Black Holes

The **Event Horizon Telescope (EHT)** captured the first-ever image of a black hole in **M87** (2019) and later **Sagittarius A*** (2022). Future telescopes will provide **sharper images and deeper insights** into black holes.

6.6.2 Searching for Wormholes

Scientists are developing **gravitational wave detectors** that could distinguish wormholes from black holes. If wormholes exist, new telescopes may reveal their presence.

6.6.3 Space Probes and Future Experiments

- **Proposals to send spacecraft near black holes** to test relativity.
- **Quantum experiments** to explore whether black holes and wormholes interact with quantum mechanics.

6.7 Conclusion: The Endless Exploration of the Unknown

Black holes and wormholes represent the **cutting edge of astrophysics**—where science meets the unknown. While we have taken great strides in understanding black holes, **wormholes and dark matter remain unsolved puzzles**. As technology advances, we may one day unlock these secrets, opening the door to new discoveries, **faster-than-light travel, and deeper knowledge of the cosmos**.

The universe is vast, mysterious, and waiting to be explored.

Chapter 7: How to Start Your Own Stargazing Hobby

7.1 Introduction: The Joy of Stargazing

Stargazing is one of the most rewarding and accessible hobbies that connects people with the vast universe. Whether you're gazing up at the stars with your naked eye, using binoculars, or exploring deep-sky objects with a telescope, the night sky offers endless wonders. Unlike many hobbies, stargazing requires little initial investment and can be enjoyed from almost anywhere on Earth.

This chapter will provide a **comprehensive guide** to starting your own **stargazing hobby**, covering **essential equipment, best observing locations, night sky navigation, astrophotography, and tips for enhancing your experience**.

7.2 Understanding the Basics of Stargazing

7.2.1 What Can You See in the Night Sky?

Depending on your location, time of year, and level of equipment, you can observe:

- **The Moon** – Craters, maria (lunar seas), and phases.
- **Planets** – Including Jupiter's moons, Saturn's rings, and Mars' surface.
- **Stars and Constellations** – The backbone of skywatching.
- **Meteor Showers** – Annual displays of shooting stars.
- **Deep-Sky Objects (DSOs)** – Nebulae, galaxies, and star clusters.

7.2.2 Why Stargazing Is Worth It

- **Educational** – Enhances knowledge of astronomy, physics, and history.
- **Relaxing and Meditative** – Provides a peaceful connection with nature.
- **Affordable** – Can be done with little to no equipment.
- **Great for All Ages** – A perfect activity for individuals, families, and communities.

7.3 Choosing the Right Equipment

7.3.1 Observing With the Naked Eye

Before investing in equipment, begin by **learning the night sky** with just your eyes. You can:

- Identify **constellations and asterisms** (e.g., Orion, the Big Dipper).
- Track **planetary movements** and bright celestial objects.
- Observe **the phases of the Moon**.
- Watch for **meteors, eclipses, and the Milky Way**.

7.3.2 Binoculars: The Beginner's Best Tool

Binoculars provide **a low-cost way** to enhance stargazing. They allow you to:

- Observe **the craters of the Moon** in detail.
- Spot **Jupiter's four largest moons**.
- See **star clusters** such as the **Pleiades (M45)**.
- Identify **nebulae and bright galaxies**, such as Andromeda.

Choosing the Right Binoculars

- **7x50**: Ideal for handheld stargazing (7x magnification, 50mm lenses).
- **10x50**: Offers more magnification, but requires a steady hand.
- **15x70+**: Best with a tripod for stability.

7.3.3 Telescopes: Expanding Your View

A telescope opens up **deeper views of space**, allowing you to see:

- **Saturn's rings** and **Jupiter's Great Red Spot**.
- **The Orion Nebula (M42)** and **Andromeda Galaxy (M31)**.
- **Lunar details** such as mountain ranges and valleys.

Types of Telescopes

- **Refractor Telescopes** – Great for planets and the Moon; minimal maintenance.
- **Reflector Telescopes** – Larger aperture for deep-sky viewing but requires collimation.
- **Compound (Catadioptric) Telescopes** – A mix of both, offering versatility but at a higher cost.

Recommended Starter Telescopes

- **Dobsonian (6"-10" aperture)** – Best for beginners due to ease of use and affordability.
- **Maksutov-Cassegrain (90mm-127mm)** – Compact and excellent for planetary viewing.
- **Go-To Telescopes** – Computerized systems that automatically locate celestial objects.

7.4 Finding the Best Stargazing Locations

7.4.1 Avoiding Light Pollution

Light pollution is one of the biggest obstacles to stargazing. The best way to experience **a dark sky** is by moving away from city lights.

- **Urban Areas** – Stick to bright objects like the Moon, planets, and major constellations.
- **Suburban Areas** – Find parks or nature reserves to reduce light interference.
- **Rural Areas** – Best for deep-sky objects, meteors, and the Milky Way.

Finding Dark Sky Locations

- Use websites like **Dark Sky Finder** or **LightPollutionMap.info**.
- National parks and remote locations offer the **best conditions**.
- Some regions are certified **Dark Sky Reserves**, ensuring low artificial light.

7.4.2 Ideal Observing Conditions

For the best stargazing experience, consider:

- **Moon Phase** – A **new moon** is best for deep-sky objects; a full moon washes out fainter stars.
- **Clear Skies** – Check weather forecasts and avoid cloudy or humid nights.
- **Stable Atmosphere** – Low turbulence improves planet and star visibility.

7.5 Learning to Navigate the Night Sky

7.5.1 Using Star Charts and Astronomy Apps

- **Printed Star Charts** – Old-school but reliable.
- **Mobile Apps** – Apps like **SkySafari, Stellarium, and SkyView** help identify celestial objects in real time.
- **Planispheres** – Rotating star charts that show the night sky based on date and time.

7.5.2 Understanding Celestial Motion

- **The Earth Rotates** – Stars appear to move westward over time.
- **Annual Changes** – Different constellations are visible in different seasons.
- **The Ecliptic** – The path where the Sun, Moon, and planets move through the sky.

7.5.3 Recognizing Constellations and Asterisms

Some of the most recognizable star patterns include:

- **The Big Dipper (Ursa Major)** – Points to Polaris, the North Star.
- **Orion (The Hunter)** – A winter constellation with bright stars **Betelgeuse and Rigel**.
- **The Summer Triangle** – Formed by **Vega, Deneb, and Altair**.

7.6 Getting Started with Astrophotography

If you want to **capture** the beauty of the night sky, astrophotography is a rewarding extension of stargazing.

7.6.1 Basic Equipment

- **Smartphone + Tripod** – Good for Moon photography and star trails.
- **DSLR Camera** – Long-exposure images of the Milky Way and constellations.
- **Telescope with a Camera Adapter** – For planetary and deep-sky imaging.

7.6.2 Best Beginner Photography Targets

- The **Moon** – Easiest and most detailed celestial object.
- **Jupiter and Saturn** – Best captured with a telescope.
- **The Milky Way** – Requires a dark-sky location and long-exposure settings.

7.7 Joining a Stargazing Community

7.7.1 Astronomy Clubs and Events

- Local astronomy clubs often host **stargazing nights** and **workshops**.
- **Star Parties** allow you to meet experienced astronomers and try different telescopes.

7.7.2 Online Astronomy Communities

- **Forums** – Cloudy Nights, AstroBin, and Reddit's r/Astronomy.
- **Live Webcasts** – NASA and observatories often stream celestial events.

7.8 Conclusion: Your Journey Begins

Starting a stargazing hobby is one of the most **accessible, enjoyable, and awe-inspiring** ways to explore the universe. With minimal equipment, the right conditions, and some guidance, anyone can embark on a **lifelong journey of celestial discovery**.

As you progress, you may find yourself drawn deeper into astronomy—whether through **telescope observations, astrophotography, or even citizen science projects**.

Look up, explore, and let the stars guide you on a journey that never ends.

Chapter 8: Space Exploration: NASA, SpaceX, and Beyond

8.1 Introduction: The Age of Space Exploration

The quest to explore space is one of humanity's greatest achievements. From the first artificial satellite to the dream of colonizing Mars, space exploration has evolved from a Cold War competition into a global, collaborative effort driven by both government agencies and private companies.

Organizations like **NASA (National Aeronautics and Space Administration)** have long been at the forefront of space exploration, leading historic missions like the **Apollo Moon landings** and the **Mars rover expeditions**. However, the 21st century has witnessed a **space revolution** led by private companies such as **SpaceX, Blue Origin, and Rocket Lab**, promising cheaper, faster, and more ambitious space missions.

This chapter explores the history, present, and future of space exploration, highlighting key missions, technological advancements, and the role of emerging space agencies and companies.

8.2 The History of Space Exploration

8.2.1 The Space Race: The Birth of Modern Spaceflight (1957-1975)

The **Space Race** was a Cold War-era competition between the **United States (NASA)** and the **Soviet Union (Roscosmos)** to achieve spaceflight milestones. Key achievements included:

- **Sputnik 1 (1957, USSR)** – The first artificial satellite, marking the start of the space age.
- **Vostok 1 (1961, USSR)** – Cosmonaut **Yuri Gagarin** became the first human in space.
- **Apollo 11 (1969, USA)** – Astronaut **Neil Armstrong** became the first human to walk on the Moon.
- **Salyut & Skylab Space Stations (1970s, USSR & USA)** – The first steps toward long-term human presence in space.

8.2.2 The Shuttle Era and the ISS (1981-Present)

The **Space Shuttle program (1981-2011)** revolutionized spaceflight with **reusable spacecraft**. Key achievements:

- **Hubble Space Telescope (1990)** – Provided groundbreaking images of deep space.
- **International Space Station (1998-present)** – A global collaboration enabling long-term human presence in orbit.

However, the **Columbia (2003) and Challenger (1986) disasters** revealed the risks of reusable spacecraft, leading to the program's retirement in 2011.

8.3 NASA: Leading the Charge in Space Exploration

NASA remains a **global leader** in space exploration, focusing on:

8.3.1 The Artemis Program: The Return to the Moon

- **Goal**: Establish a **permanent lunar base** to prepare for Mars missions.
- **Artemis I (2022)**: An uncrewed mission testing the **Orion spacecraft** and the **Space Launch System (SLS)**.
- **Artemis II (2024-2025)**: First crewed lunar flyby since Apollo.
- **Artemis III (2026)**: The first humans on the Moon since 1972.

8.3.2 Mars Exploration: The Search for Life

- **Perseverance Rover (2021)**: Searching for signs of past microbial life.
- **Ingenuity Helicopter**: The first controlled flight on another planet.
- **Mars Sample Return (2030s)**: A mission to bring Martian soil back to Earth.

8.3.3 Deep Space Missions

- **James Webb Space Telescope (2021-Present)**: The most powerful space observatory, capable of detecting distant galaxies and exoplanets.
- **Europa Clipper (2025 Launch)**: Exploring Jupiter's moon Europa, which may harbor an ocean beneath its icy surface.
- **Dragonfly (2027 Launch)**: A drone mission to Saturn's moon **Titan**, known for its methane lakes.

8.4 SpaceX: The Commercial Space Revolution

Founded in 2002 by **Elon Musk**, SpaceX has **transformed space travel** by making rockets reusable and significantly reducing launch costs.

8.4.1 Key Achievements of SpaceX

- **Falcon 1 (2008)** – First privately developed liquid-fueled rocket to reach orbit.
- **Falcon 9 (2015-Present)** – First reusable orbital-class rocket.
- **Dragon Capsule (2012-Present)** – First private spacecraft to dock with the ISS.
- **Crew Dragon (2020)** – First private company to send astronauts to the ISS.

8.4.2 Starship: The Future of Space Travel

SpaceX is developing **Starship**, a fully reusable spacecraft designed for:

- **Lunar Landings** (Artemis Program).
- **Mars Colonization** (Musk's goal: "Making life multiplanetary").
- **Interplanetary Cargo Transport** (Delivering large payloads across the Solar System).

If successful, **Starship will be the most powerful rocket ever built**, capable of carrying **100+ passengers to Mars**.

8.5 Other Key Players in Space Exploration

8.5.1 Blue Origin: Jeff Bezos' Space Venture

Founded by **Amazon's Jeff Bezos**, Blue Origin focuses on:

- **Suborbital Tourism** – **New Shepard** takes passengers briefly into space.
- **Orbital Missions** – **New Glenn** will rival SpaceX's Falcon 9.
- **Lunar Exploration** – The **Blue Moon lander** is designed for NASA's Artemis program.

8.5.2 Other Emerging Space Companies

- **Rocket Lab** – Specializes in small satellite launches.
- **Boeing** – Developing the **Starliner crew capsule**.
- **Virgin Galactic** – Offers **suborbital space tourism**.

8.6 The Future of Space Exploration

8.6.1 The International Space Station's Successor

- The ISS will be **retired in 2030**, with **private space stations** planned as replacements.
- NASA is funding **Axiom Space** to develop **commercial orbital habitats.**

8.6.2 Space Tourism and Commercialization

- Companies like **SpaceX, Blue Origin, and Virgin Galactic** are making space accessible to civilians.
- Future plans include **space hotels** and **lunar tourism**.

8.6.3 Mars Colonization: A Realistic Goal?

- **SpaceX aims to land humans on Mars by the 2030s**.
- Challenges include **radiation exposure, life support, and food production**.
- The first Mars settlers may **live underground** to shield from radiation.

8.6.4 Interstellar Travel: Science Fiction or Future Reality?

- **Breakthrough Starshot**: A project aiming to send small **light-powered spacecraft to Alpha Centauri**.
- **Nuclear Fusion Propulsion**: A potential game-changer for deep space missions.
- **Wormholes & Warp Drives**: Still purely theoretical but explored in advanced physics.

8.7 Ethical and Legal Considerations of Space Exploration

8.7.1 Who Owns Space?

- The **Outer Space Treaty (1967)** states that no nation can claim celestial bodies.
- However, private companies are now pushing for **resource rights** on the Moon and asteroids.

8.7.2 Space Debris and Environmental Concerns

- **Thousands of satellites** orbit Earth, creating a **growing space junk problem**.
- Solutions include **active debris removal** and **biodegradable satellites**.

8.7.3 Should Humans Colonize Other Worlds?

- Ethical debates continue over **terraforming Mars** and the potential contamination of alien environments.

8.8 Conclusion: A New Space Age Begins

We are living in an **unprecedented era** of space exploration, where **public agencies and private companies** work together to push humanity beyond Earth.

- **NASA's Artemis Program** aims to **return to the Moon** and prepare for Mars.
- **SpaceX and Blue Origin** are making space more **accessible, affordable, and sustainable**.
- **Future technologies** could take us to **Mars, exoplanets, and beyond**.

The **next 50 years** will determine whether humanity becomes **a true spacefaring civilization**. The journey has begun, and the stars await.

Look up, dream big, and explore—because the universe is ours to discover.

Chapter 9: The Search for Extraterrestrial Life

9.1 Introduction: Are We Alone in the Universe?

For centuries, humanity has looked up at the night sky and wondered if we are alone. The search for extraterrestrial life is one of the most profound scientific endeavors, exploring the possibility that life exists beyond Earth. This search extends beyond simple curiosity; discovering life elsewhere would redefine our understanding of biology, evolution, and our place in the cosmos.

With advances in astronomy, planetary science, and astrobiology, scientists are closer than ever to finding evidence of life beyond our planet. This chapter delves into the **scientific methods, technological advancements, and potential locations** where alien life may exist, from microbial organisms on Mars to intelligent civilizations in distant galaxies.

9.2 Defining Life Beyond Earth

Before searching for extraterrestrial life, we must define what life is. On Earth, life is characterized by:

- **Cellular structure** (made of organic molecules).
- **Metabolism** (converting energy for survival).
- **Reproduction** (passing genetic material to offspring).
- **Adaptation** (evolution through natural selection).

Extraterrestrial life may be **radically different** from Earth's biology, possibly existing in environments unlike anything we have seen. Scientists classify potential alien life into:

1. **Microbial Life (Simple Life Forms)** – Bacteria-like organisms that might thrive under extreme conditions.
2. **Complex Multicellular Life** – Organisms similar to plants, fungi, or animals.
3. **Intelligent Life** – Civilizations capable of communication and technology.

9.3 The Search for Life in Our Solar System

9.3.1 Mars: The Most Likely Candidate

Mars has been the focus of astrobiological research for decades due to its similarities to early Earth. Key discoveries include:

- **Liquid Water Evidence** – Seasonal water flows detected on Mars' surface.
- **Organic Molecules** – Curiosity Rover found complex carbon compounds.
- **Methane Fluctuations** – Methane spikes suggest potential biological or geological activity.

Future missions, such as **NASA's Mars Sample Return (2030s)**, aim to bring Martian soil back to Earth for analysis.

9.3.2 Europa: Jupiter's Ocean World

Europa, one of Jupiter's largest moons, is covered in an icy shell, beneath which lies a **global liquid water ocean**. Scientists speculate that:

- Hydrothermal vents at the ocean floor could provide conditions similar to Earth's deep-sea ecosystems.
- Plumes of water vapor erupting from Europa's surface may contain organic molecules.

NASA's **Europa Clipper (2025 Launch)** will analyze Europa's surface and subsurface for potential life.

9.3.3 Enceladus: Saturn's Active Moon

Similar to Europa, Saturn's moon **Enceladus** has:

- A subsurface ocean with organic material.
- Water plumes shooting into space, making it easy to collect samples.
- Possible hydrothermal activity, which could support microbial life.

The **Dragonfly mission (2027)** will explore Saturn's moon **Titan**, which has lakes of liquid methane and ethane, potentially harboring exotic life forms.

9.4 The Search for Exoplanets and Habitable Worlds

9.4.1 The Habitable Zone: The Goldilocks Region

The **habitable zone** is the region around a star where conditions may allow for liquid water, a key ingredient for life.

Factors that make a planet potentially habitable:

- **Right distance from the star** (not too hot or cold).
- **Atmospheric conditions** that can support life.
- **Presence of water and essential elements** like carbon, nitrogen, and oxygen.

9.4.2 Exoplanet Discoveries

Over **5,000 exoplanets** have been discovered, with some resembling Earth. Key discoveries include:

- **Proxima b** – A planet in the habitable zone of **Proxima Centauri**, 4.2 light-years away.
- **TRAPPIST-1 System** – A star system with **seven Earth-sized planets**, three of which may have liquid water.
- **Kepler-442b** – One of the most Earth-like exoplanets, with conditions favorable for life.

The **James Webb Space Telescope (JWST)** is analyzing exoplanet atmospheres for biosignatures, such as **oxygen, methane, and water vapor**.

9.5 The Search for Intelligent Life: The Fermi Paradox and SETI

9.5.1 The Fermi Paradox: Where is Everyone?

The universe is **13.8 billion years old**, with billions of potentially habitable planets. If intelligent life is common, why haven't we found it yet? Possible explanations include:

- **The Great Filter** – Civilizations may destroy themselves before becoming interstellar.
- **We're Too Early or Too Late** – Intelligent life may not have emerged yet, or it already went extinct.
- **They Are Avoiding Us** – Advanced civilizations may not wish to interact with primitive life forms.
- **They Exist, But We Can't Detect Them** – Alien signals might be undetectable with current technology.

9.5.2 The Search for Extraterrestrial Intelligence (SETI)

SETI uses **radio telescopes** to scan the sky for signals from alien civilizations.

Notable attempts:

- **Wow! Signal (1977)** – A mysterious 72-second radio signal detected from space, never repeated.
- **Breakthrough Listen (2015-Present)** – A $100 million initiative scanning nearby stars for alien transmissions.

Future technologies, such as **AI-based signal processing**, could improve our ability to detect alien messages.

9.6 Alternative Theories About Alien Life

9.6.1 Could Aliens Be Completely Different from Life on Earth?

Life elsewhere may not be based on **carbon and water**. Alternative forms could include:

- **Silicon-based life** (alternative to carbon-based organisms).
- **Plasma-based life** (existence in high-energy environments).
- **Methane-based life** (potentially thriving on Titan).

9.6.2 UFOs and the Possibility of Extraterrestrial Visitors

Governments and space agencies have declassified UFO (Unidentified Aerial Phenomena, UAP) reports. Some scientists remain open to the idea that certain unexplained sightings **could be advanced alien probes.**

9.6.3 Panspermia: Did Life on Earth Come from Space?

Some scientists propose that **life originated elsewhere and was transported to Earth** via:

- **Meteorites** containing amino acids.
- **Interstellar comets** spreading microbial life across planets.

If true, life may be **more widespread in the universe** than we thought.

9.7 The Future of the Search for Alien Life

9.7.1 Upcoming Missions and Telescopes

- **James Webb Space Telescope (JWST)** – Scanning exoplanet atmospheres for life signatures.
- **Europa Clipper (2025)** – Investigating Europa's ocean for potential microbes.
- **Mars Sample Return (2030s)** – Bringing Martian soil back to Earth for testing.
- **Breakthrough Starshot (2050s)** – Sending nanocrafts to the nearest exoplanets.

9.7.2 What If We Discover Alien Life?

A discovery of extraterrestrial life—whether microbial or intelligent—would have **profound implications**:

- **Religious and philosophical debates** on our place in the cosmos.
- **Scientific breakthroughs** in biology and evolution.
- **Potential contact protocols** and ethical considerations.

Would we attempt to communicate? Or should we stay silent to avoid drawing attention to ourselves?

9.8 Conclusion: A Universe Full of Possibilities

The search for extraterrestrial life is one of the most exciting frontiers of modern science. From microbial life on Mars to intelligent civilizations across the galaxy, the possibilities are endless. As technology advances, we may be on the verge of answering the most profound question:

Are we alone in the universe?

The next few decades **could change everything**. Until then, we continue to listen, explore, and look to the stars, hoping to find our cosmic neighbors.

The truth is out there—waiting to be discovered.

Chapter 10: The Expanding Universe and Dark Matter

10.1 Introduction: The Hidden Forces Shaping the Cosmos

The universe is not static; it is **expanding at an accelerating rate.** This fundamental discovery reshaped our understanding of cosmology and raised profound questions about the **fate of the universe** and the **invisible forces that dominate it.**

One of these forces is **dark matter**, an unknown substance that holds galaxies together, yet remains undetectable by traditional means. Another is **dark energy**, a mysterious force driving the universe's accelerated expansion. Together, these components make up **95% of the universe's total composition**, yet their nature remains one of the greatest unsolved mysteries in modern physics.

This chapter explores the **expansion of the universe, the evidence for dark matter, and the role of dark energy**, providing a deep dive into the cutting-edge research that seeks to explain these cosmic enigmas.

10.2 The Expanding Universe: A Revolutionary Discovery

10.2.1 Hubble's Law and the Expanding Universe

The concept of an expanding universe was first proposed in the early 20th century. Before this, scientists believed the universe was eternal and unchanging. However, in the **1920s**, astronomer **Edwin Hubble** made a groundbreaking discovery:

- By observing distant galaxies, Hubble noticed that their **light was redshifted**—a sign that they were moving **away from us**.
- He formulated **Hubble's Law**, which states that **the farther away a galaxy is, the faster it is receding**.

This led to the realization that **space itself is expanding**, much like a balloon inflating.

10.2.2 The Big Bang Theory

The **Big Bang Theory** is the leading explanation for the origin of the universe. Key ideas include:

- **13.8 billion years ago**, the universe was an incredibly hot, dense point.
- A **rapid expansion** (inflation) caused space to stretch and cool.
- Over time, particles formed atoms, galaxies, and stars.

The **cosmic microwave background (CMB)**—a faint glow detected across the universe—is the leftover radiation from the Big Bang, providing strong evidence for this theory.

10.2.3 Cosmic Inflation: What Happened Before the Big Bang?

Physicist **Alan Guth** proposed **cosmic inflation**, suggesting that in the first **fractions of a second**, the universe expanded faster than the speed of light. This explains:

- **The uniformity of the universe** (why galaxies are evenly spread out).
- **The tiny fluctuations that led to galaxy formation**.

However, what triggered inflation—and what came **before the Big Bang**—remains unknown.

10.3 The Role of Dark Matter: The Hidden Mass of the Universe

10.3.1 What is Dark Matter?

Dark matter is an invisible substance that:

- **Does not emit, absorb, or reflect light**, making it impossible to detect directly.
- Exerts **gravitational effects**, holding galaxies together.
- Makes up **27% of the universe's total mass-energy composition**.

10.3.2 Evidence for Dark Matter

Even though we can't see dark matter, its effects are evident through:

1. **Galaxy Rotation Curves**
    - Stars at the edges of galaxies orbit **just as fast** as those near the center.
    - According to Newtonian physics, outer stars should move slower—unless **extra mass (dark matter) is present**.
2. **Gravitational Lensing**
    - Large clusters of galaxies bend and distort light from objects behind them.
    - The amount of bending suggests **more mass than visible matter can account for**.
3. **The Cosmic Web Structure**
    - Dark matter provides the **scaffolding** upon which galaxies and clusters form.
    - Without it, the large-scale structure of the universe would not exist as we see it today.

10.3.3 What is Dark Matter Made Of?

Scientists propose several candidates for dark matter:

- **WIMPs (Weakly Interacting Massive Particles)** – Hypothetical particles that interact only through gravity and the weak nuclear force.
- **Axions** – Extremely light particles that could explain some quantum mysteries.
- **Primordial Black Holes** – Small black holes formed after the Big Bang.

Experiments like the **Large Hadron Collider (LHC)** and underground detectors are actively searching for dark matter particles.

10.4 Dark Energy: The Force Driving Expansion

10.4.1 Discovery of Dark Energy

In **1998**, astronomers analyzing distant **supernovae** discovered that the universe's expansion was not slowing down as expected—it was **accelerating**.

To explain this, scientists proposed **dark energy**, an unknown force making up **68% of the universe**.

10.4.2 Theories Behind Dark Energy

Several theories attempt to explain dark energy:

1. **The Cosmological Constant (Λ)**
    - Proposed by **Albert Einstein** in his equations of General Relativity.
    - A constant energy inherent in space itself, causing expansion.
2. **Quintessence**
    - A dynamic energy field that evolves over time.
3. **Modified Gravity Theories**
    - Suggest that Einstein's theory of gravity is incomplete on cosmic scales.

10.4.3 The Fate of the Universe: How Will It End?

The expansion of the universe raises questions about its ultimate fate:

1. **The Big Freeze (Heat Death)** – If dark energy continues accelerating expansion, galaxies will **drift apart**, stars will burn out, and the universe will end in **cold darkness**.
2. **The Big Rip** – Dark energy **increases in strength**, tearing apart galaxies, atoms, and even spacetime itself.
3. **The Big Crunch** – If dark energy weakens, gravity could **reverse expansion**, causing the universe to collapse back into a singularity.

Currently, **the Big Freeze is the most likely scenario**, based on observational data.

10.5 How Scientists Study the Expanding Universe and Dark Matter

10.5.1 Telescopes and Observatories

1. **Hubble Space Telescope** – Provided deep-space images that confirmed accelerated expansion.
2. **James Webb Space Telescope (JWST)** – Studying the earliest galaxies to understand cosmic evolution.
3. **Vera C. Rubin Observatory (2025 launch)** – Will map dark matter using gravitational lensing.

10.5.2 Particle Physics and Dark Matter Detection

1. **Large Hadron Collider (LHC)** – Searching for unknown particles that could be dark matter.
2. **XENON Experiment (Italy)** – Detecting possible dark matter interactions underground.
3. **NASA's Euclid Mission (2023 launch)** – Mapping dark matter and dark energy effects on the universe's structure.

10.6 Unsolved Mysteries and Future Research

Despite decades of research, dark matter and dark energy remain unsolved puzzles.

Key unanswered questions:

- What exactly is **dark matter**, and why can't we detect it?
- Is dark energy a **property of space** or a **new fundamental force**?
- Could there be **hidden dimensions** influencing cosmic expansion?

Future missions, such as the **Nancy Grace Roman Space Telescope (2027 launch)**, aim to shed light on these mysteries.

10.7 Conclusion: The Quest to Understand the Universe

The **expansion of the universe** and the **mysteries of dark matter and dark energy** represent the cutting edge of astrophysics. These invisible forces **shape the cosmos**, yet remain beyond our grasp.

As telescopes grow more powerful and particle experiments reach new energy levels, we may one day uncover the truth about these enigmatic cosmic forces.

Until then, the universe remains an **unfolding mystery**, inviting us to keep looking, questioning, and exploring.

The answers are out there—waiting to be discovered.

Appendix A: Best Telescopes and Apps for Stargazing

A.1 Introduction: Essential Tools for Stargazing

Stargazing is one of the most accessible and rewarding hobbies, whether you are using just your **naked eye, binoculars, or a telescope**. With advancements in technology, amateur astronomers now have access to **powerful telescopes and mobile apps** that make identifying celestial objects easier than ever.

This appendix serves as a **comprehensive guide** to choosing the best telescopes for beginners and advanced users, along with the **top astronomy apps** that can enhance your stargazing experience.

A.2 Best Telescopes for Stargazing

Choosing the right telescope depends on **your experience level, budget, and observing goals**. Below is a breakdown of the best telescopes for **beginners, intermediate users, and advanced astronomers.**

A.2.1 Telescopes for Beginners

If you are just starting, consider a telescope that is **easy to set up, user-friendly, and affordable.**

1. Celestron StarSense Explorer DX 102AZ

- **Type**: Refractor
- **Aperture**: 102mm
- **Best For**: Planets, Moon, bright deep-sky objects
- **Price**: $$
- **Why It's Great**: Uses **StarSense smartphone technology** to automatically align and locate celestial objects.

2. Orion SkyScanner 100mm Tabletop Reflector

- **Type**: Reflector
- **Aperture**: 100mm
- **Best For**: Moon, planets, some bright deep-sky objects
- **Price**: $
- **Why It's Great**: Small, portable, and provides **clear views of Jupiter, Saturn, and the Moon.**

3. Celestron AstroMaster 70AZ

- **Type**: Refractor
- **Aperture**: 70mm
- **Best For**: Moon, planets
- **Price**: $
- **Why It's Great**: Affordable, lightweight, and great for beginners.

A.2.2 Intermediate Telescopes

For those who want **deeper sky views**, these telescopes offer **larger apertures and better optics.**

4. Orion SkyQuest XT8 Classic Dobsonian

- **Type**: Reflector (Dobsonian)
- **Aperture**: 203mm (8 inches)
- **Best For**: Deep-sky objects (galaxies, nebulae), planets
- **Price**: $$$
- **Why It's Great**: **Best value aperture** for deep-sky viewing. Dobsonians provide **bright, clear images**.

5. Celestron NexStar 6SE

- **Type**: Schmidt-Cassegrain
- **Aperture**: 150mm (6 inches)
- **Best For**: Planets, nebulae, galaxies
- **Price**: $$$
- **Why It's Great**: **GoTo technology** locates and tracks objects automatically.

6. Meade LX85 6" Reflector Telescope

- **Type**: Newtonian Reflector
- **Aperture**: 150mm (6 inches)
- **Best For**: Deep-sky viewing, astrophotography
- **Price**: $$$
- **Why It's Great**: Equatorial mount for **long-exposure astrophotography**.

A.2.3 Advanced Telescopes

If you are looking for **professional-level observations**, these high-end telescopes provide **superior optics and imaging capabilities**.

7. Sky-Watcher 12" Collapsible Dobsonian

- **Type**: Reflector (Dobsonian)
- **Aperture**: 305mm (12 inches)
- **Best For**: Faint galaxies, nebulae
- **Price**: $$$$
- **Why It's Great**: **Massive aperture**, collapsible design for portability.

8. Celestron CGX-L 925 EdgeHD

- **Type**: Schmidt-Cassegrain
- **Aperture**: 235mm (9.25 inches)
- **Best For**: Deep-sky astrophotography, planetary viewing
- **Price**: $$$$$
- **Why It's Great**: **Computerized equatorial mount** for precise tracking.

9. Meade 16-inch LX600 ACF Advanced Coma-Free Telescope

- **Type**: Advanced Schmidt-Cassegrain
- **Aperture**: 406mm (16 inches)
- **Best For**: Professional deep-sky imaging
- **Price**: $$$$$$
- **Why It's Great**: One of the **most powerful consumer telescopes** available.

A.3 Best Astronomy Apps for Stargazing

Modern **stargazing apps** use **augmented reality (AR), GPS tracking, and databases** to help users locate and identify celestial objects.

A.3.1 Best Free Astronomy Apps

1. Stellarium Mobile (iOS/Android)

- **Best For**: Real-time night sky navigation.
- **Features**:
    - Identifies stars, planets, and constellations.
    - Augmented Reality (AR) mode.
    - Works offline.

2. SkyView Lite (iOS/Android)

- **Best For**: Simple AR-based sky identification.
- **Features**:
    - Point your phone at the sky to identify objects.
    - Includes planets, satellites, and constellations.

3. Star Walk 2 Free (iOS/Android)

- **Best For**: Educational stargazing.
- **Features**:
    - Interactive 3D model of the sky.
    - Beautiful graphics and animations.

A.3.2 Best Paid Astronomy Apps

4. SkySafari 7 Pro ($) (iOS/Android)

- **Best For**: Advanced stargazers and telescope control.
- **Features**:
    - Largest celestial database.
    - Control GoTo telescopes.
    - Augmented reality tracking.

5. Night Sky X ($) (iOS)

- **Best For**: Premium AR stargazing experience.
- **Features**:
    - Hyper-detailed 3D models of celestial objects.
    - Notifications for upcoming space events.

6. Heavens-Above (Free/Paid) (iOS/Android)

- **Best For**: Tracking satellites, including the ISS.
- **Features**:
    - Live satellite tracking.
    - Shows visible passes of the **International Space Station (ISS)**.

A.3.3 Best Apps for Astrophotography

7. PhotoPills ($) (iOS/Android)

- **Best For**: Planning astrophotography shots.
- **Features**:
    - Helps plan Milky Way and lunar photography.
    - Calculates light pollution levels.

8. Deep Sky Camera (Android)

- **Best For**: Capturing long-exposure images.
- **Features**:
    - Stacks multiple images to enhance clarity.
    - Custom exposure settings.

A.4 Conclusion: The Right Tools for Your Stargazing Journey

Whether you are a **casual observer or an aspiring astronomer**, the right **telescope and app** can greatly enhance your experience.

- **Beginners** should start with **a simple telescope or binoculars and apps like Stellarium**.
- **Intermediate users** will benefit from **a Dobsonian telescope or GoTo systems**.
- **Advanced astronomers** should invest in **high-aperture telescopes and astrophotography tools**.

With the help of **modern technology**, exploring the universe has never been easier. No matter your level, **the night sky is yours to discover!**

Message from the Author:

I hope you enjoyed this book, I love astrology and knew there was not a book such as this out on the shelf. I love metaphysical items as well. Please check out my other books:

-Life of Government Benefits

-My life of Hell

-My life with Hydrocephalus

-Red Sky

-World Domination:Woman's rule

-World Domination:Woman's Rule 2: The War

-Life and Banishment of Apophis: book 1

-The Kidney Friendly Diet

-The Ultimate Hemp Cookbook

-Creating a Dispensary(legally)

-Cleanliness throughout life: the importance of showering from childhood to adulthood.

-Strong Roots: The Risks of Overcoddling children

-Hemp Horoscopes: Cosmic Insights and Earthly Healing

- Celestial Hemp Navigating the Zodiac: Through the Green Cosmos

-Astrological Hemp: Aligning The Stars with Earth's Ancient Herb

-The Astrological Guide to Hemp: Stars, Signs, and Sacred Leaves

-Green Growth: Innovative Marketing Strategies for your Hemp Products and Dispensary

-Cosmic Cannabis

-Astrological Munchies

-Henry The Hemp

-Zodiacal Roots: The Astrological Soul Of Hemp

- **Green Constellations: Intersection of Hemp and Zodiac**

-Hemp in The Houses: An astrological Adventure Through The Cannabis Galaxy

-Galactic Ganja Guide

Heavenly Hemp

Zodiac Leaves

Doctor Who Astrology

Cannastrology

Stellar Satvias and Cosmic Indicas

Celestial Cannabis: A Zodiac Journey

AstroHerbology: The Sky and The Soil: Volume 1

AstroHerbology:Celestial Cannabis:Volume 2

Cosmic Cannabis Cultivation

The Starry Guide to Herbal Harmony: Volume 1

The Starry Guide to Herbal Harmony: Cannabis Universe: Volume 2

Yugioh Astrology: Astrological Guide to Deck, Duels and more

Nightmare Mansion: Echoes of The Abyss

Nightmare Mansion 2: Legacy of Shadows

Nightmare Mansion 3: Shadows of the Forgotten

Nightmare Mansion 4: Echoes of the Damned

The Life and Banishment of Apophis: Book 2

Nightmare Mansion: Halls of Despair

Healing with Herb: Cannabis and Hydrocephalus

Planetary Pot: Aligning with Astrological Herbs: Volume 1

Fast Track to Freedom: 30 Days to Financial Independence Using AI, Assets, and Agile Hustles

Cosmic Hemp Pathways

How to Become Financially Free in 30 Days: 10,000 Paths to Prosperity

Zodiacal Herbage: Astrological Insights: Volume 1

Nightmare Mansion: Whispers in the Walls
The Daleks Invade Atlantis
Henry the hemp and Hydrocephalus

10X The Kidney Friendly Diet
Cannabis Universe: Adult coloring book
Hemp Astrology: The Healing Power of the Stars
Zodiacal Herbage: Astrological Insights: Cannabis Universe: Volume 2
Planetary Pot: Aligning with Astrological Herbs: Cannabis Universes: Volume 2
Doctor Who Meets the Replicators and SG-1: The Ultimate Battle for Survival
Nightmare Mansion: Curse of the Blood Moon
The Celestial Stoner: A Guide to the Zodiac
Cosmic Pleasures: Sex Toy Astrology for Every Sign
Hydrocephalus Astrology: Navigating the Stars and Healing Waters
Lapis and the Mischievous Chocolate Bar

Celestial Positions: Sexual Astrology for Every Sign
Apophis's Shadow Work Journal: **:** A Journey of Self-Discovery and Healing
Kinky Cosmos: Sexual Kink Astrology for Every Sign
Digital Cosmos: The Astrological Digimon Compendium
Stellar Seeds: The Cosmic Guide to Growing with Astrology
Apophis's Daily Gratitude Journal

Cat Astrology: Feline Mysteries of the Cosmos
The Cosmic Kama Sutra: An Astrological Guide to Sexual Positions
Unleash Your Potential: A Guided Journal Powered by AI Insights
Whispers of the Enchanted Grove

Cosmic Pleasures: An Astrological Guide to Sexual Kinks

369, 12 Manifestation Journal

Whisper of the nocturne journal(blank journal for writing or drawing)

The Boogey Book

Locked In Reflection: A Chastity Journey Through Locktober

Generating Wealth Quickly:How to Generate $100,000 in 24 Hours

Star Magic: Harness the Power of the Universe

The Flatulence Chronicles: A Fart Journal for Self-Discovery

The Doctor and The Death Moth

Seize the Day: A Personal Seizure Tracking Journal

The Ultimate Boogeyman Safari: A Journey into the Boogie World and Beyond

Whispers of Samhain: 1,000 Spells of Love, Luck, and Lunar Magic: Samhain Spell Book

Apophis's guides:Witch's Spellbook Crafting Guide for Halloween

Frost & Flame: The Enchanted Yule Grimoire of 1000 Winter Spells

The Ultimate Boogey Goo Guide & Spooky Activities for Halloween Fun

Harmony of the Scales: A Libra's Spellcraft for Balance and Beauty

The Enchanted Advent: 36 Days of Christmas Wonders

Nightmare Mansion: The Labyrinth of Screams

Harvest of Enchantment: 1,000 Spells of Gratitude, Love, and Fortune for Thanksgiving

The Boogey Chronicles: A Journal of Nightly Encounters and Shadowy Secrets

The 12 Days of Financial Freedom: A Step-by-Step Christmas Countdown to Transform Your Finances

Sigil of the Eternal Spiral Blank Journal

A Christmas Feast: Timeless Recipes for Every Meal

Holiday Stress-Free Solutions: A Survival Guide to Thriving During the Festive Season

Yu-Gi-Oh! Holiday Gifting Mastery: The Ultimate Guide for Fans and Newcomers Alike

Holiday Harmony: A Hydrocephalus Survival Guide for the Festive Season

Celestial Craft: The Witch's Almanac for 2025 – A Cosmic Guide to Manifestations, Moons, and Mystical Events

Doctor Who: The Toymaker's Winter Wonderland

Tulsa King Unveiled: A Thrilling Guide to Stallone's Mafia Masterpiece

Pendulum Craft: A Complete Guide to Crafting and Using Personalized Divination Tools

Nightmare Mansion: Santa's Eternal Eve

Starlight Noel: A Cosmic Journey through Christmas Mysteries

The Dark Architect: Unlocking the Blueprint of Existence

Surviving the Embrace: The Ultimate Guide to Encounters with The Hugging Molly

The Enchanted Codex: Secrets of the Craft for Witches, Wiccans, and Pagans

Harvest of Gratitude: A Complete Thanksgiving Guide

Yuletide Essentials: A Complete Guide to an Authentic and Magical Christmas

Celestial Smokes: A Cosmic Guide to Cigars and Astrology

Living in Balance: A Comprehensive Survival Guide to Thriving with Diabetes Insipidus

Cosmic Symbiosis: The Venom Zodiac Chronicles

The Cursed Paw of Ambition

Cosmic Symbiosis: The Astrological Venom Journal

Celestial Wonders Unfold: A Stargazer's Guide to the Cosmos (2024-2029)

The Ultimate Black Friday Prepper's Guide: Mastering Shopping Strategies and Savings

Cosmic Sales: The Astrological Guide to Black Friday Shopping

Legends of the Corn Mother and Other Harvest Myths

Whispers of the Harvest: The Corn Mother's Journal

The Evergreen Spellbook

The Doctor Meets the Boogeyman

The White Witch of Rose Hall's SpellBook

The Gingerbread Golem's Shadow: A Study in Sweet Darkness

The Gingerbread Golem Codex: An Academic Exploration of Sweet Myths

The Gingerbread Golem Grimoire: Sweet Magicks and Spells for the Festive Witch

The Curse of the Gingerbread Golem

10-minute Christmas Crafts for kids

Christmas Crisis Solutions: The Ultimate Last-Minute Survival Guide

Gingerbread Golem Recipes: Holiday Treats with a Magical Twist

The Infinite Key: Unlocking Mystical Secrets of the Ages

Enchanted Yule: A Wiccan and Pagan Guide to a Magical and Memorable Season

Dinosaurs of Power: Unlocking Ancient Magick

Astro-Dinos: The Cosmic Guide to Prehistoric Wisdom

Gallifrey's Yule Logs: A Festive Doctor Who Cookbook

The Dino Grimoire: Secrets of Prehistoric Magick

The Gift They Never Knew They Needed

The Gingerbread Golem's Culinary Alchemy: Enchanting Recipes for a Sweetly Dark Feast

A Time Lord Christmas: Holiday Adventures with the Doctor

Krampusproofing Your Home: Defensive Strategies for Yule

Silent Frights: A Collection of Christmas Creepypastas to Chill Your Bones

Santa Raptor's Jolly Carnage: A Dino-Claus Christmas Tale

Prehistoric Palettes: A Dino Wicca Coloring Journey
The Christmas Wishkeeper Chronicles
The Starlight Sleigh: A Holiday Journey
Elf Secrets: The True Magic of the North Pole
Candy Cane Conjurations
Cooking with Kids: Recipes Under 20 Minutes
Doctor Who: The TARDIS Confiscation
The Anxiety First Aid Kit: Quick Tools to Calm Your Mind
Frosty Whispers: A Winter's Tale
The Infinite Key: Unlocking the Secrets to Prosperity, Resilience, and Purpose
The Grasping Void: Why You'll Regret This Purchase
Astrology for Busy Bees: Star Signs Simplified
The Instant Focus Formula: Cut Through the Noise
The Secret Language of Colors: Unlocking the Emotional Codes
Sacred Fossil Chronicles: Blank Journal
The Christmas Cottage Miracle
Feeding Frenzy: Graboid-Inspired Recipes
Manifest in Minutes: The Quick Law of Attraction Guide
The Symbiote Chronicles: Doctor Who's Venomous Journey
Think Tiny, Grow Big: The Minimalist Mindset
The Energy Key: Unlocking Limitless Motivation
New Year, New Magic: Manifesting Your Best Year Yet
Unstoppable You: Mastering Confidence in Minutes
Infinite Energy: The Secret to Never Feeling Drained
Lightning Focus: Mastering the Art of Productivity in a Distracted World
Saturnalia Manifestation Magick: A Guide to Unlocking Abundance During the Solstice
Graboids and Garland: The Ultimate Tremors-Themed Christmas Guide
12 Nights of Holiday Magic
The Power of Pause: 60-Second Mindfulness Practices

The Quick Reset: How to Reclaim Your Life After Burnout

The Shadow Eater: A Tale of Despair and Survival

The Micro-Mastery Method: Transform Your Skills in Just Minutes a Day

Reclaiming Time: How to Live More by Doing Less

Chronovore: The Eternal Nexus

The Mind Reset: Unlocking Your Inner Peace in a Chaotic World

Confidence Code: Building Unshakable Self-Belief

Baby the Vampire Terrier

Baby the Vampire Terrier's Christmas Adventure

Celestial Streams: The Content Creator's Astrology Manual

The Wealth Whisperer: Unlocking Abundance with Everyday Actions

The Energy Equation: Maximize Your Output Without Burning Out

The Happiness Algorithm: Science-Backed Steps to Joyful Living

Stress-Free Success: Achieving Goals Without Anxiety

Mindful Wealth: The New Blueprint for Financial Freedom

The Festive Flavors of New Year: A Culinary Celebration

The Master's Gambit: Keys of Eternal Power

Shadowed Secrets: Groundhog Day Mysteries

Beneath the Burrow: Lessons from the Groundhog

Spring's Whispers: The Groundhog's Prediction

The Limitless Mindset: Unlock Your Untapped Potential

The Focus Funnel: How to Cut Through Chaos and Get Results

Bold Moves: Building Courage to Live on Your Terms

The Daily Shift: Simple Practices for Lasting Transformation

The Quarter-Life Reset: Thriving in Your 20s and 30s

The Art of Shadowplay: Building Your Own Personal Myth

The Eternal Loop: Finding Purpose in Repetition

Burrowing Wisdom: Life Lessons from the Groundhog

Shadow Work: A Groundhog Day Perspective

Love in Bloom: 5-Minute Romantic Gestures

The Shadowspell Codex: Secrets of Forbidden Magick
The Burnout Cure: Finding Balance in a Busy World
The Groundhog Prophecy: Unlocking Seasonal Secrets
Nog Tales: The Spirited History of Eggnog
Six More Weeks: Embracing Seasonal Transitions
The Lumivian Chronicles: Fragments of the Fifth Dimension
Money on Your Mind: A Beginner's Guide to Wealth
The Focus Fix: Breaking Through Distraction
January's Spirit Keepers: Mystical Protectors of the Cold
Creativity Unchained: Unlocking Your Wildest Ideas in 2025
Manifestation Mastery: 365 Days to Rewrite Your Reality
The Groundhog's Mirror: Reflecting on Change
The Weeping Angels' Christmas Curse
Burrowed in Time: A Groundhog Day Journey
Heartbeats: Poems to Share with Your Valentine
Dino Wicca: The Sacred Grimoire of Prehistoric Magick
Courage of the Pride: Finding Your Inner Roar
The Lion's Leap: Bold Moves for Big Results
Healthy Hustle: Achieving Without Overworking
Practical Manifesting: Turning Dreams into Reality in 2025
Jurassic Pharaohs: Unlocking the Magick of Ancient Egypt and Dino Wicca
The Happiness Equation: Small Changes for Big Joy
The Confidence Compass: Finding Your Inner Strength
Whispers in the Hollow: Tales of the Forgotten Beasts
Echoes from the Hollow: The Return of Forgotten Beasts
The Hollow Ascendant: The Rise of the Forgotten Beasts
The Relationship Reset: Building Better Connections
Mastering the Morning: How to Win the Day Before 8 AM
The Shadow's Dance: Groundhog Day Symbolism
Cupid's Kitchen: Quick Valentine's Day Recipes
Valentine's Day on a Budget: Love Without Breaking the Bank
Astrocraft: Aligning the Stars in the World of Minecraft

Forecasting Life: Groundhog Day Reflections

Bleeding Hearts: Twisted Tales of Valentine's Terror

Herbal Smoke Revolution: The Ultimate Guide to Nature's Cigarette Alternative

Winter's Wrath: The Complete Survival Blueprint for Extreme Freezes.

The Groundhog's Shadow: A Tale of Seasons

Burrowed Insights: Wisdom from the Groundhog

Sensual Strings: The Art of Erotic Bondage

Whispered Flames: Unlocking the Power of Fire Play

Forgotten Shadows: A Guide to Cryptids Lost to Time

Six Weeks of Secrets: Groundhog Day's Hidden Messages

Shadows and Cycles: Groundhog Day Reflections

The Art of Love Letters: Crafting the Perfect Message

Romantic Getaways at Home: Turning Your Space into Paradise

Purrfect Brews: A Cat Lover's Guide to Coffee and Companionship

The Groundhog's Wisdom: Timeless Lessons for Modern Life

The Shadow Oracle: Groundhog Day as a Predictor

Emerging from the Burrow: A Journey of Renewal

The Language of Love: Learning Your Partner's Love Style

Authorpreneur: The Ultimate Blueprint for Writing, Publishing, and Thriving as an Author

Weathering the Seasons: Groundhog Day Perspectives

Valentine's Day Magic: A Guide to Romantic Rituals

The Shadow Chronicles: Stories of Groundhog Day

Love and Laughter: Fun Games for Valentine's Day

AstroRealty: Unlocking the Stars for Property Success

The Groundhog's Path: A Guide to Seasonal Balance

Groundhog Day Diaries: Reflections in the Shadow

The Groundhog's Light: Illuminating the Path Ahead

Valentine's Traditions from Around the World

AI Wealth Revolution: Unlocking the Trillionaire Mindset
Love Rekindled: Reigniting Passion in Relationships
Single and Thriving: Self-Love on Valentine's Day
Emerald Legends: Mystical Tales of Ireland
Green Alchemy: Harnessing Nature's Magic
The Hearts of Horror: A Valentine's Day Nightmare
The Leprechaun's Guide to Wealth and Wisdom
Dancing with the Sidhe: Celebrating the Otherworld
Shamrocks and Shadows: Mysteries of the Green Isle
Emerald Energy: Harnessing Luck and Growth
The Gingerbread Golem's Valentine: A Sweetheart's Guide to Love and Enchantment
The Celtic Knot: Weaving Life and Destiny
Green Fire: Elemental Magic for St. Patrick's Day
Clover Chronicles: Finding Your Inner Luck
Ireland's Mystical Creatures: A Field Guide
Gingerbread Golem's Love Almanac
Prowl and Thrive: The Lion's Guide to Success
Love Alchemy: Transforming Your Life Through Heart Energy
WORLD DOMINATION: Woman's Rule 3:The New Life
The Midnight Rose: A Guide to Lunar Love Spells
The Forbidden Letters: Writing Your Own Love Prophecy
Luck and Lore: St. Patrick's Day for Modern Mystics
The Green Path: A Pagan Celebration of Renewal
The Dark Architect's Guide to Reprogramming Reality
Prankster's Paradise: A Guide to Harmless Hijinks
Manifest Your Reality: The Law of Attraction Simplified
The TARDIS Owner's Manual: Understanding the Doctor's Ship: *A complete guide to the TARDIS, its technology, secrets, and mysteries*
Starlit Romance: Astrology Secrets for Finding True Love

The Time Lord's Atlas: A Complete Guide to the Whoniverse: *A breakdown of the locations, planets, and dimensions explored in Doctor Who*

Sweetheart Shadows: The Dark Side of Love and Attraction

February Fire: Reigniting Passion in Every Area of Life

The Self-Love Toolkit: 5 Ways to Embrace Who You Are

February Sparks: Ignite Your Dreams in 28 Days

March to Success: A 31-Day Action Blueprint

Ancient Paths: The 13 Sacred Principles of Dino Wicca

Echoes of Tomorrow: Navigating the AI Revolution

The Wellness Blueprint: Balancing Mind, Body, and Soul

Green Horizons: Sustainable Living for a Better Tomorrow

The AI Wealth Code: How to Make Millions with Automation

AI-Powered Creativity: Writing, Art, and Music for Profit

Extinction Rites: Rebirthing Your Soul Through Prehistoric Magick

Sacred Serpents tarot

Celestial Enchantment blank journal

Star Strains

Culinary Journeys: Exploring Global Flavors at Home

The Hollowvale Curse

The Hollowvale Harvest

The Egg of Transformation: Awakening Your Inner Power

Blooming Into Power: A Wiccan Guide to Spring Awakening

The Nightmare Nexus: The Third Doctor's Perilous Haunting

Digital Detox: Reclaiming Your Life in a Connected World

Ostara's Path: Walking the Spiral of Renewal

The Sacred Hare

Financial Freedom: Building Wealth in the Modern Age

Spring's Cauldron: Stirring the Waters of Change

The Hollowvale Pact

Quantum Consciousness: The Science of Reality Shifting

The Hollowvale Hunger

The Sacred Waters Within: A Witch's Guide to Hydrocephalus Magick

The Raven's Nest: Building a Life of Unshakable Stability

AI and the Human Mind: The Future of Intelligence

The Hollowvale Reckoning

Timeless Love: Building and Maintaining Lasting Relationships

The Raven's Roar: Unlocking Unstoppable Confidence

Raven Sight: Awakening Intuition and Inner Wisdom

The Butterfly Effect: Small Changes, Big Transformations

Taming the Boogeyman: How to Conquer Your Inner Fears

The Magick of Green: Awakening Earth's Energy in You

The Entrepreneurial Mindset: Secrets to Business Success

The Hollowvale End

The Shadow Luck Ritual: Reclaiming Power from Your Dark Side

Spring Magick for Beginners: A Simple Guide to Seasonal Energy Work

Doctor Who: The Hollowvale Conundrum

The March of Miracles: Unlocking Synchronicities in Spring

If you want solar for your home go here: https://www.harborsolar.live/apophisenterprises/

Get Some Tarot cards: https://www.makeplayingcards.com/sell/apophis-occult-shop

Instagrams:
@apophis_enterprises,
@apophisbookemporium,
@apophisscardshop
Twitter: @apophisenterpr1
Tiktok:@apophisenterprise
Youtube: @sg1fan23477, @FiresideRetreatKingdom
Hive: @sg1fan23477
CheeLee: @SG1fan23477

Podcast: Apophis Chat Zone: https://open.spotify.com/show/5zXbrCLEV2xzCp8ybrfHsk?si=fb4d4fdbdce44dec

Newsletter: https://apophiss-newsletter-27c897.beehiiv.com/

If you want to support me or see posts of other projects that I have come over to: **buymeacoffee.com/mpetchinskg**

I post there daily several times a day

Get your Dinowicca or Christmas themed digital products, especially Santa Raptor songs and other musics. Here: **https://sg1fan23477.gumroad.com**

Apophis Yuletide Digital has not only digital Christmas items, but it will have all things with Dinowicca as well as other Digital products.

www.ingramcontent.com/pod-product-compliance
Lightning Source LLC
LaVergne TN
LVHW022007200926
848481LV00007B/2271
* 9 7 9 8 3 4 8 5 8 3 5 9 0 *